YOU WILL PASS TRIGONOMETRY

You Will Pass Trigonometry

Poetry Affirmations for Math Students

Walter the Educator™

SKB

Silent King Books a WhichHead Imprint

Copyright © 2023 by Walter the Educator™

All rights reserved. No part of this book may be reproduced in any manner whatsoever without written permission except in the case of brief quotations embodied in critical articles and reviews.

First Printing, 2023

Disclaimer
This book is a literary work; poems are not about specific persons, locations, situations, and/or circumstances unless mentioned in a historical context. This book is for entertainment and informational purposes only. The author and publisher offer this information without warranties expressed or implied. No matter the grounds, neither the author nor the publisher will be accountable for any losses, injuries, or other damages caused by the reader's use of this book. The use of this book acknowledges an understanding and acceptance of this disclaimer.

dedicated to all the math lovers across the world

CONTENTS

Dedication v

Why I Created This Book? 1

One - Beauty Of Trigonometry 2

Two - Fear Not 4

Three - Embrace The Grace 6

Four - You'll Find The Power 8

Five - Discover The Secrets 10

Six - Within The Struggle 12

Seven - Transforming Challenges 14

Eight - Mathematical Flight 16

Nine - Reach For The Sky 18

Ten - Your Passion Ignite 20

Eleven - Conquer The Obstacles 22

Twelve - Power Lies Within You 24

Thirteen - Unleashing Powers	. . . .	26
Fourteen - Passing Trigonometry	. . . .	28
Fifteen - Success Will Be Your Name	. . . .	30
Sixteen - Believe In Yourself	. . . .	32
Seventeen - Trials And Tribulations	. . . . .	34
Eighteen - Dear Student	. . . .	36
Nineteen - Pass This Subject	. . . .	38
Twenty - Let Your Dreams Rise	. . . .	40
Twenty-One - Future Shines Bright	. . . . . .	42
Twenty-Two - Triangles As Your Allies	. . . .	44
Twenty-Three - Trigonometry's Song	. . . . .	46
Twenty-Four - Make It Your Own	. . . .	48
Twenty-Five - Resolve To Evolve	. . . .	50
Twenty-Six - Journey To Success	. . . .	52
Twenty-Seven - Stay The Course	. . . .	54
Twenty-Eight - Your Brilliance Shine	. . . . .	56
Twenty-Nine - Soar Through The Heavens	. .	58
Thirty - Dive Deep	. . . .	60
Thirty-One - Success Will Surely Come To You	. . . .	62

Thirty-Two - Strong And True 64

Thirty-Three - Determination As Your Loyal Friend 66

Thirty-Four - Take A Leap 68

Thirty-Five - Trigonometry Can Set You Free 70

About The Author 72

WHY I CREATED THIS BOOK?

Creating a poetry book to motivate students to pass the subject of Trigonometry was a unique and effective way to engage and inspire them. Poetry has the power to convey complex ideas and emotions in a creative and memorable manner. By using poetic language, metaphors, and imagery, this book can make the subject of Trigonometry more relatable and accessible to students who may find it challenging or uninteresting. Additionally, this book can instill a sense of confidence, perseverance, and determination in students, encouraging them to overcome difficulties and succeed in their studies.

ONE

BEAUTY OF TRIGONOMETRY

In the realm of numbers and angles, behold,
Trigonometry, a subject bold.
With triangles and ratios, it does entwine,
A journey of discovery, so divine.

Oh student, listen to my heartfelt plea,
Unlock the secrets, set your knowledge free.
Embrace the challenge, let curiosity guide,
For in Trigonometry, wonders reside.

When angles dance and lines converge,
Precision and logic, let them emerge.
Delve into the depths of this mathematical sea,
And unlock the mysteries that await thee.

Fear not the complexities that may arise,
For with determination, you shall rise.

Oh student, let not frustration consume,
For perseverance shall be your ultimate bloom.
    Seek the patterns that lie within,
Connect the dots, let the learning begin.
With every theorem and formula you grasp,
Confidence in Trigonometry shall clasp.
    Remember, dear student, the power you hold,
To conquer this subject, brave and bold.
Let passion ignite, let knowledge soar,
And Trigonometry's triumph, you shall explore.
    So, march forward, with courage and might,
Illuminate the path, banish the night.
For in your hands, lies the key,
To unlock the beauty of Trigonometry.

TWO

FEAR NOT

In the realm of numbers, where angles reside,
Lies a subject waiting, with secrets to confide.
Trigonometry, a puzzle to unravel,
A journey of discovery, where minds can travel.
    Oh student, embrace this challenge with might,
Let not frustration dim your passionate light.
For within these formulas, a beauty unfolds,
A symphony of patterns, waiting to be told.
    Unlock the wonders of triangles, my dear,
Let your curiosity be your guiding spear.
Seek the connections, the dots to align,
And watch as the mysteries gracefully intertwine.
    Persevere, young scholar, through every trial,
For knowledge and wisdom will be your smile.

Let not the complexities weigh down your soul,
With determination, you can reach your goal.

With each equation conquered, you'll soar high,
A winged mathematician, touching the sky.
Illuminate the path with your power and might,
And witness the marvels hidden in plain sight.

Trigonometry, a gateway to understanding,
A subject that sets your intellect expanding.
With courage as your ally, you shall surpass,
And find solace in the wonders of this class.

So, fear not, dear student, let your passion ignite,
For Trigonometry reveals its beauty, day and night.
Unlock the mysteries, let your knowledge soar,
And embrace the challenges that lie at its core.

THREE

EMBRACE THE GRACE

In the realm of angles and curves, behold,
A subject of wonder, yet to unfold.
Trigonometry, a mystical art,
That beckons you to play your part.
    Amidst the theorems and equations so grand,
Unveil the secrets of this sacred land.
Let not frustration cloud your sight,
But bask in the glory of trigonometric light.
    In the depths of triangles, you will find,
A universe of patterns, intertwined.
Like stars in the night, they dance and align,
Unlocking the mysteries, one by one, sublime.
    Embrace the challenge, let passion ignite,
For within this subject, lies infinite delight.

With determination as your guiding star,
You'll conquer the complexities, no matter how far.
   Let curiosity be your faithful guide,
As you journey through this mathematical tide.
For in the depths of its elegant embrace,
Lie the wonders that only few can trace.
   Embrace the beauty, let it be your muse,
For within its grasp, knowledge shall infuse.
Trigonometry holds the key to your expansion,
To understanding and intellectual comprehension.
   So, fear not the numbers, the angles, or the lines,
For within them, a symphony of brilliance shines.
In your hands lies the power to pass this test,
To conquer the subject and be truly blessed.
   Let your courage soar, let your spirit be free,
For Trigonometry is where you're meant to be.
Embrace the challenge, embrace the grace,
And triumph in this wondrous mathematical space.

FOUR

YOU'LL FIND THE POWER

In the realm of angles and lines,
Where mathematics intertwines,
Lies a subject, both feared and revered,
Trigonometry, to be endeared.

Fear not, dear student, the challenge you face,
For within these numbers, a hidden grace.
Unlock the secrets, embrace the unknown,
Let Trigonometry be your very own.

From right triangles to radians and more,
Discover the beauty you can't ignore.
In graphs and formulas, a world unfolds,
Where patterns and wonders gently behold.

Let curiosity guide your quest,
As you delve into this mystical nest.

Unearth the mysteries, one by one,
And witness the magic that can't be undone.

    Trigonometry, a journey profound,
Where logic and artistry are tightly bound.
So fear not, dear student, for you hold the key,
To conquer this subject with bravery.

    Let passion ignite, let courage prevail,
As you navigate through each intricate trail.
With determination, you'll rise above,
And unlock the vastness of Trigonometry's love.

    For in its embrace, you'll find the power,
To conquer complexities, hour by hour.
Embrace the subject, let it expand your sight,
And soar through the heavens, like a star shining bright.

FIVE

DISCOVER THE SECRETS

Oh, student, do not falter in thy quest,
To conquer Trigonometry's noble test.
Embrace the challenge with a heart so bold,
And watch thy knowledge, like a star, unfold.
  For in the realm of angles and triangles,
Lies a world of wonders, endless like the seas.
Discover the secrets of sine, cosine, and tan,
Unlock the beauty of this sacred plan.
  From right triangles to circles so grand,
Trigonometry paints a world by thy hand.
It weaves connections, unseen and profound,
Like a symphony, harmonious in sound.
  In the depths of equations and ratios untold,
Lies the path to wisdom, a treasure to behold.

Fear not the numbers that dance in the air,
For they hold the key to dreams beyond compare.
So, let not despair cloud thy bright mind,
For within Trigonometry, solace thou shall find.
With each triumph, a victory so sweet,
A testament to thy resilience, so complete.
Rise, dear student, with courage in thy stride,
Let Trigonometry be thy faithful guide.
Pass this subject with unfailing might,
And unlock a world of endless delight.

SIX

WITHIN THE STRUGGLE

Oh, brave student, do not shy away,
For Trigonometry's mysteries hold sway,
In angles and triangles, secrets unfold,
A universe of knowledge waiting to be told.

Through the depths of sin and cos,
Lie the answers that you seek, close,
With perseverance and unwavering might,
You can conquer the challenges, take flight.

Trigonometry, a puzzle to unravel,
But fear not, for you are strong and able,
In the realm of triangles, you'll find,
A world of wonder, so vast and kind.

Let curiosity guide your way,
And the beauty of Trigonometry will sway,

Its intricate patterns and elegant design,
Will captivate your mind, oh so fine.

    Embrace the angles, embrace the lines,
For in their dance, true understanding shines,
Let not frustration cloud your view,
For within the struggle lies the breakthrough.

    So fear not the equations, fear not the test,
For you are capable, you are blessed,
Unlock the wonders that Trigonometry holds,
And find solace and delight in its folds.

    With every step, with every try,
You'll triumph over Trigonometry's sky,
Embrace the challenge, embrace the quest,
And in your success, you'll find true zest.

SEVEN

TRANSFORMING CHALLENGES

In the realm of angles and shapes,
Lies a subject that often escapes,
The grasp of many, causing distress,
But fear not, dear student, I must confess.

Trigonometry, a world of its own,
With complexities that can make you groan,
But within its depths, wonders unfold,
A treasure trove waiting to be behold.

Let not frustration dim your sight,
Embrace the challenge, with all your might,
For within the numbers and equations galore,
Lies a path to wisdom, forevermore.

Triangles dance with elegance and grace,
Their secrets waiting for you to embrace,

Sine, cosine, tangent, and more,
Unlocking their mysteries, you'll surely adore.

With compass in hand, navigate the unknown,
Discovering patterns, like a melody's tone,
Let curiosity guide your way,
And Trigonometry's secrets will sway.

So, fear not the theorems that lie ahead,
For with determination, you'll forge ahead,
Embrace the beauty of this ancient art,
And Trigonometry's triumphs will impart.

Passion and perseverance, your guiding light,
Transforming challenges into pure delight,
With every step, you'll rise above,
And conquer Trigonometry with unwavering love.

So, dear student, let your spirit soar,
In the realm of Trigonometry, forever explore,
For in your journey, knowledge will abide,
And with it, success will be your guide.

EIGHT

MATHEMATICAL FLIGHT

In the realm where angles dance with glee,
Lies a path to unlock the mystery.
Trigonometry, a world untamed,
With numbers and shapes to be proclaimed.

    Fear not, dear student, for within your grasp,
Lies the power to conquer and unclasp,
The secrets of triangles, both big and small,
And navigate through curves, standing tall.

    Let curiosity be your guiding light,
As you delve into this mathematical flight.
Embrace the challenge, let passion ignite,
And watch your understanding take flight.

    For Trigonometry's beauty lies in its might,
To measure the heavens, the stars so bright.

From the depths of the ocean to the highest peak,
It weaves its magic, the truth it seeks.
    With formulas and theorems, you'll navigate,
Through angles and ratios, never abate.
With every step, your knowledge will grow,
And the wonders of Trigonometry will surely show.
    So fear not the tests, the equations, or sums,
For within you lies the power to overcome.
With determination and a curious mind,
Trigonometry's mysteries, you shall find.
    Pass this subject, unlock its door,
And soar to heights you've never soared before.
For in conquering Trigonometry's domain,
You'll reap the rewards, forever sustain.

NINE

REACH FOR THE SKY

In the realm of numbers and angles, behold,
A subject waits, its mysteries untold.
Trigonometry, a path less trodden,
Where brilliance shines, and minds are sodden.

    Oh student, fear not the tests and equations,
For within lies the beauty of transformations.
Embrace the triangles, their angles so true,
And let curiosity guide you through.

    In the dance of sines and cosines, you'll find,
A world of patterns, intricately designed.
Let determination be your guiding light,
As you unravel the secrets with all your might.

    For in Trigonometry, there lies a key,
Unlocking the door to knowledge's decree.

With every step, you'll rise above the rest,
And conquer the challenges, be at your best.
  Let not frustration hinder your way,
For perseverance will lead you astray.
With every hurdle, you'll grow strong,
And prove to the world that you belong.
  For Trigonometry holds the power to ignite,
A passion within, burning ever so bright.
So let your spirit soar, reach for the sky,
And watch as your dreams take flight, oh so high.
  In this realm of triangles and equations,
A world of possibilities awaits your creations.
Pass Trigonometry, and you shall see,
The boundless potential that lies within thee.

TEN

YOUR PASSION IGNITE

In the realm of angles and circles we tread,
Lies a subject that fills many minds with dread,
Trigonometry, with its complex equations and sums,
But fear not, dear student, for the adventure has just begun.

Let curiosity guide you through its mystical maze,
With perseverance, you'll unravel its dazzling ways,
For in the depths of Trigonometry's embrace,
Lies a world of knowledge, waiting to grace.

With triangles and ratios, we'll dance with delight,
As we uncover the secrets hidden in plain sight,
Angles and sine, cosine, and tangent's embrace,
Will lead us to triumph, at a steady, determined pace.

So fear not the tests that lie in your path,
Embrace the challenge, let your spirit soar and laugh,

For Trigonometry's beauty lies not in its ease,
But in the transformation it brings, if you please.

Unlock the power of triangles, with heart and with will,
And watch as your mind expands, and dreams fulfill,
With each step forward, a new horizon is revealed,
Trigonometry, the gateway to a world unconcealed.

So, dear student, let your passion ignite,
For in Trigonometry, you'll find your true height,
Believe in yourself, and the wonders you'll see,
As you conquer the subject, and set your spirit free.

ELEVEN

CONQUER THE OBSTACLES

In the realm of numbers and angles, behold,
A subject awaits, with its secrets untold.
Trigonometry, a path to explore,
Where knowledge and beauty forever endure.

Oh, student, fear not this mystical art,
For within its equations lies wisdom's heart.
Let curiosity guide your eager mind,
And the wonders of triangles you shall find.

Angles and ratios, they dance in accord,
Revealing the mysteries that lie untoward.
With sine and cosine, the stars align,
Unlocking the universe, oh, so divine.

Through triangles and circles, patterns emerge,
A language of shapes, where insights converge.

From Pythagoras' theorem to the unit circle's grace,
Trigonometry unveils the world's true embrace.

In this journey, challenges may arise,
But with perseverance, you'll reach the skies.
Embrace the formulas, with determination strong,
And conquer the obstacles that come along.

For in the realm of Trigonometry's land,
Lies the power to transform and expand.
With every theorem learned and problem solved,
Your potential and success will surely evolve.

So, march forth, dear student, with passion and might,
Let Trigonometry be your guiding light.
Embrace the beauty, the challenges, the rewards,
And soar to new heights, where knowledge affords.

TWELVE

POWER LIES WITHIN YOU

In the realm of angles, where numbers dance,
Lies the enigma of Trigonometry's trance.
A subject mysterious, yet beautifully divine,
A gateway to knowledge, secrets to find.
   Embrace the triangles, my eager friend,
For in their depths, wisdom transcends.
Let curiosity fuel your quest,
And Trigonometry will reveal its best.
   With sine and cosine, tangent and more,
Unlock the wonders you've yet to explore.
From right-angled triangles to unit circles' grace,
Trigonometry unveils its stunning embrace.
   Amidst the formulas and identities to discover,
Lies the power to conquer, to soar and recover.

Though challenges may arise on this journey you tread,
With perseverance, you'll conquer the dread.
    No longer a puzzle, but a pathway to see,
The interconnectedness of math's vast decree.
Unlock the secrets of stars, waves, and space,
With Trigonometry as your guiding trace.
    Embrace the challenge, let passion ignite,
And watch as your understanding takes flight.
For in the realm of Trigonometry's art,
Lies the potential for growth, a brand new start.
    Believe in yourself, let doubt be erased,
For Trigonometry's beauty should never be misplaced.
Pass the test, but remember this, too,
Trigonometry's true power lies within you.

THIRTEEN

UNLEASHING POWERS

In the realm of angles and lines,
Where Trigonometry truly shines,
A student embarks on a quest,
To conquer the subject, be their best.

With triangles and ratios to explore,
Their determination shall soar,
Trigonometry, a puzzle to unravel,
A journey that will surely marvel.

Embrace the challenge, do not fear,
For knowledge awaits when you draw near,
Angles and degrees, they hold the key,
To unlock the beauty, for all to see.

Let the stars guide your way,
As you navigate through the array,
Cosines and sines, they intertwine,
Unleashing powers, divine and fine.

In this realm of calculations and arcs,
You'll discover the wonders and sparks,
Trigonometry, a gateway to dreams,
Where possibilities gleam and gleam.

Have faith in yourself, believe you can,
Reach for the skies, become a fan,
For Trigonometry reveals the grand,
The symphony of numbers, a magical strand.

So, student, don't falter or despair,
Embrace Trigonometry with flair,
In its depths, you'll find your might,
And pass this subject, shining bright.

FOURTEEN

PASSING TRIGONOMETRY

In the realm of angles and triangles,
Where numbers dance and equations rise,
Lies the power of Trigonometry,
Unleashing the potential within our eyes.

Oh, weary student, do not despair,
For Trigonometry is a path to embrace,
A gateway to a universe so fair,
Where logic and beauty interlace.

With every sine and cosine you trace,
The secrets of the universe unfold,
Unlocking the mysteries of time and space,
Revealing treasures yet untold.

Let not the complexities hinder your stride,
For challenges are what make you grow,

Embrace the journey with arms open wide,
And watch your knowledge bloom and flow.

For in the realm of Trigonometry,
Lies the key to understanding it all,
The power to shape your destiny,
And rise above the shadows so small.

Believe in yourself, oh student dear,
Let not doubts dim your radiant light,
With passion and perseverance, have no fear,
For Trigonometry will guide you right.

So, grasp the stars within your reach,
And let your dreams take flight,
For in passing Trigonometry, you'll teach,
The world that you're a shining light.

FIFTEEN

SUCCESS WILL BE YOUR NAME

In the realm of angles and triangles,
Lies a subject that often brings sighs.
Trigonometry, a mysterious art,
Where numbers and shapes play their part.

At first, it may seem daunting and complex,
But fear not, for you're up to the test.
Unlock the secrets of sine and cosine,
And watch as your knowledge begins to shine.

Trigonometry, a gateway to the stars,
Where you'll chart celestial bodies afar.
Measure the heavens with angles precise,
And see the beauty in their cosmic dance.

From waves that ripple to sound in the air,
Trigonometry reveals patterns so rare.

Explore the depths of sound and light,
And marvel at the wonders within your sight.
   Harness the power of triangles' might,
And conquer the challenges with all your might.
For in Trigonometry's realm you'll find,
A world of possibilities, vast and kind.
   So persevere through the trials you face,
And let your determination embrace.
For in the realm of Trigonometry,
Lies the key to unlock your true destiny.
   Believe in yourself, let your potential ignite,
And soar to new heights, like a radiant light.
Trigonometry's rewards are yours to claim,
With passion and perseverance, success will be your name.

SIXTEEN

BELIEVE IN YOURSELF

In the realm of angles and lines,
Where numbers dance and stars align,
Lies the subject that may confound,
But fear not, for you are bound.

    Trigonometry, a cosmic maze,
A puzzle of celestial ways,
Unlock the secrets it conceals,
And witness the wonders it reveals.

    With triangles as your guiding light,
You'll navigate through day and night,
From sine to cosine, let them be,
The compass to set your spirit free.

    Embrace the challenge, don't shy away,
For greatness lies in your display,

In graphs and formulas, you'll find,
A universe waiting to unwind.
 Through radians and degrees you'll tread,
A journey where your dreams are led,
In every curve and every line,
Your potential shines, oh so divine.
 So let your passion be your guide,
And in Trigonometry, take pride,
For in its depths, you will find,
The power to shape your own design.
 Believe in yourself, oh student dear,
For Trigonometry holds the key,
To unlock the doors of destiny,
And let your brilliance truly appear.

SEVENTEEN

TRIALS AND TRIBULATIONS

In the realm of numbers and angles, a story unfolds,
Trigonometry, a journey of wonders, untold.
With triangles as our guide, we navigate the unknown,
Unlocking the secrets of the universe, we're shown.

Within these lines and curves, lies a hidden power,
A language of harmony, in every shape and tower.
Embrace the challenge, let your spirit soar,
For Trigonometry's beauty, we cannot ignore.

From sine to cosine, and tangent's embrace,
We dance with the stars, in this cosmic space.
Let the equations guide you, like a compass so true,
And the mysteries of the cosmos will be unveiled to you.

Through the trials and tribulations, you'll find your

way,
With every problem solved, you'll grow stronger each day.
Believe in yourself, for you hold the key,
To unlock the boundless potential that lies within thee.

So, fear not the numbers, nor the formulas you face,
For they are but stepping stones, to a brighter place.
In the realm of Trigonometry, you'll discover your might,
And shape your own destiny, with knowledge and light.

Passing this subject is not just a test,
But a testament to your brilliance, your very best.
So, embrace the challenge, let your passion ignite,
And soar to new heights, with Trigonometry as your flight.

EIGHTEEN

DEAR STUDENT

In the realm of numbers, let us embark,
On a journey of angles, light, and arc.
Trigonometry, a world so vast,
Where equations dance, and concepts last.

Fear not the triangles, for they hold the key,
To unravel the secrets of geometry.
With sine, cosine, and tangent in hand,
We'll navigate this mathematical land.

Through right angles and hypotenuse,
We'll find the strength to break through,
The fear and doubt that cloud our mind,
And leave our worries far behind.

For Trigonometry is more than just math,
It's a gateway to knowledge, a creative path.

In its depths lie the mysteries of sound and light,
From harmonies in music to the stars shining bright.

    So let us embrace this subject with zeal,
Let determination be our driving wheel.
With practice and perseverance, we shall prevail,
And see our efforts and hard work unveil.

    For in Trigonometry, we find our own worth,
A chance to grow, a chance for rebirth.
Believe in yourself, dear student, and take flight,
For Trigonometry will guide you to new heights.

NINETEEN

PASS THIS SUBJECT

In the realm of numbers, where angles reside,
Lies a subject, Trigonometry, a thrilling ride.
Fear not, dear student, for it holds the key,
To unlock the mysteries of the cosmic sea.

Embrace the challenge, let your spirit soar,
Trigonometry's beauty, you shall explore.
From triangles' secrets to circles' embrace,
Let its elegance and power leave you amazed.

Like a musician, playing notes in harmony,
Trigonometry weaves patterns with such artistry.
Sines and cosines, waves in the night,
Unravel the rhythm, bring melodies to light.

In the realm of stars, where galaxies dance,
Trigonometry reveals their cosmic romance.
Measure the heavens, map the unknown,
Discover the wonders that lie beyond our own.

Believe in yourself, let confidence arise,
In Trigonometry's realm, doubts shall demise.
In every challenge, an opportunity awaits,
To grow, to learn, to elevate your fates.

So, let not fear nor doubt hold you back,
Embrace Trigonometry, stay on the right track.
With passion and perseverance, you shall thrive,
And conquer the subject, come out alive.

For in the realm of numbers, where knowledge resides,
Trigonometry beckons you to new heights.
Unlock your potential, let your brilliance shine,
Pass this subject, and a world of possibilities will be thine.

TWENTY

LET YOUR DREAMS RISE

In the realm of angles and lines,
Where triangles dance and intertwine,
Lies the path to knowledge divine,
Trigonometry, a subject so fine.

Fear not the numbers, the sines and cos,
For within each problem, a strength will grow,
Unlocking doors to a world unknown,
Where possibilities endlessly flow.

Embrace the challenge with open arms,
For in struggle, greatness is often born.
Through the depths of triangles, you'll find,
A universe of wisdom, one of a kind.

Let not frustration hinder your quest,
For every hurdle, you'll be truly blessed.

In perseverance lies the key,
To conquer Trigonometry, and be set free.
 The stars themselves, they use these laws,
To navigate the cosmos without pause,
So too shall you, with knowledge profound,
Shape your destiny, forever unbound.
 Believe in yourself, let doubt be gone,
For within your heart, the strength is strong.
Pass Trigonometry, reach for the skies,
Unleash your potential, let your dreams rise.

TWENTY-ONE

FUTURE SHINES BRIGHT

In the realm of angles, where numbers dance,
Trigonometry unveils its cosmic stance.
A subject of beauty, both abstract and real,
Where triangles whisper secrets, concealed.

Let us embark on this journey anew,
Where formulas and theorems will guide you through.
Fear not the sin, the cos, or the tan,
For in their embrace, your success will expand.

From the heights of mountains to the depths of the sea,
Trigonometry unveils the world's mystery.
It maps the stars, their patterns and light,
Guiding sailors through oceans, day and night.

In music's symphony, it finds its place,

The rhythm of angles, harmonies embrace.
The dance of chords, the melody's sway,
Trigonometry, the maestro's display.

Embrace the challenge, let your fears dissipate,
For in perseverance, you'll unlock the gate.
Believe in yourself, in your potential to soar,
And Trigonometry's secrets, you will explore.

So, student, take heart and heed my call,
Trigonometry's lessons will never enthrall.
Unlock your mind, let knowledge take flight,
For in conquering this subject, your future shines bright.

Pass the test, let success be your guide,
With Trigonometry, you'll reach new heights.
Embrace the challenge, let your dreams unfurl,
For in mastering this subject, you'll change the world.

TWENTY-TWO

TRIANGLES AS YOUR ALLIES

In the realm of angles and lines,
Where the mysteries of shapes combine,
Lies Trigonometry, a subject to embrace,
A challenge that holds your dreams in space.

Like a compass guiding you through the night,
Trigonometry unveils patterns, shining bright,
Unlocking the secrets of triangles and more,
Revealing the beauty you've yet to explore.

With triangles as your allies, you shall see,
The power of Trigonometry, set yourself free,
For in this world of numbers and ratios,
Lies a gateway to life's endless possibilities.

Embrace the challenge, fear not the unknown,
For within lies the seeds of knowledge sown,

With persistence and dedication, you will find,
The strength to conquer, to leave no doubts behind.

    Trigonometry is a key to unravel the unknown,
A path to success, a foundation firmly sown,
So rise, dear student, face the heights,
And soar through the skies, embracing the lights.

    For in Trigonometry, you'll find your voice,
A power to conquer, to rejoice,
With every formula, every theorem you learn,
You'll see your potential, your dreams return.

    So let not the fears of angles deter,
Embrace Trigonometry, for it shall infer,
That within you lies the strength and might,
To conquer the subject, to shine so bright.

TWENTY-THREE

TRIGONOMETRY'S SONG

In the realm of numbers, where angles reside,
There lies a subject, often feared with stride.
Trigonometry, the gateway to the skies,
Unlocks the secrets that make stars arise.

Oh student, do not fear this path unknown,
For in its depths, your brilliance will be shown.
Embrace the challenge, let your spirit soar,
And watch the mysteries of the universe unfold.

With triangles and circles, you'll navigate,
Through sine and cosine, your way illuminate.
Each problem solved, a victory profound,
A testament to the knowledge you have found.

From right triangles to spherical delights,
Trigonometry unveils celestial sights.

The planets dance, the constellations gleam,
And you, dear student, will be part of the dream.
 So persevere, believe in your own might,
And conquer Trigonometry's sacred height.
Unlock your potential, let your brilliance bloom,
For in this subject, the cosmos finds its room.
 Pass not for a grade, but for the joy it brings,
For Trigonometry's song within your heart sings.
Embrace the challenge, let your spirit shine,
And behold the wonders of this cosmic design.

TWENTY-FOUR

MAKE IT YOUR OWN

In the realm of angles and triangles,
A world of numbers and cosmic mysteries,
Lies the path to unlock the skies,
Trigonometry, where greatness lies.

Embrace the challenge, don't shy away,
For within its depths, knowledge holds sway.
Angles and ratios, they may seem complex,
But persevere, and you'll outshine the rest.

Let your determination be your guiding light,
As you traverse this mathematical flight.
With every concept you comprehend,
A new horizon, your mind will transcend.

For Trigonometry, the language of the stars,
Reveals the secrets of the universe afar.

From celestial bodies to the tides of the sea,
Its power and beauty, for all to see.
So, hold your compass, and grasp your pen,
As you journey through this mathematical den.
With every problem you solve and equation you derive,
You'll find the strength within you to thrive.
Believe in yourself, dear student of mine,
Through Trigonometry, your potential will shine.
Unlock the doors to a world unknown,
And conquer the subject, make it your own.

TWENTY-FIVE

RESOLVE TO EVOLVE

In the realm of numbers and angles,
Where Trigonometry's beauty entangles,
A student's journey begins anew,
To conquer the challenges, both old and new.
   Oh, dear student, fear not this path,
For within lies the power to unlock your math.
Embrace the triangles, their secrets unfold,
A cosmic dance of numbers, yet to be told.
   Let not frustration dim your flame,
For in perseverance lies fortune and fame.
With every equation, every problem you face,
You navigate the cosmos, leaving your trace.
   Like a shooting star across the night,
You'll soar through the depths, shining bright.

Trigonometry's language, a celestial song,
Guiding you through the challenges, lifelong.
From the heights of mountains to the depths of the sea,
Trigonometry unveils the world's mystery.
Architects and astronomers, they understand,
The power of triangles, the universe in hand.
So, dear student, let your doubts disperse,
Embrace the challenge, immerse in the universe.
For within the realm of Trigonometry's art,
Lies the key to reach the stars, to play your part.
With every step you take, with every problem you solve,
You'll find the strength within, the resolve to evolve.
Believe in yourself, let your passion ignite,
And Trigonometry's secrets, you shall ignite.

TWENTY-SIX

JOURNEY TO SUCCESS

In the realm of angles and curves,
Where numbers dance and knowledge swerves,
Lies a subject, both feared and revered,
Trigonometry, it's time to be endeared.

Oh student, with weary eyes and a heavy sigh,
Fear not the triangles that mystify,
For within their shapes lies a hidden key,
To unlock the wonders of the vast geometry.

Embrace the challenge, don't shy away,
Let perseverance guide you through the way,
From sine to cosine, and tangent too,
Discover the secrets that await for you.

Through radians and degrees, let your mind soar,
Explore the patterns, forever wanting more,

For Trigonometry is not a foe,
But a gateway to knowledge, don't you know?
    So study hard and don't lose faith,
In this subject that tests your mental state,
For when you conquer the mysteries untold,
A world of possibilities you shall behold.
    With every equation you learn to solve,
A new pathway in your mind evolves,
And as you grasp the concepts tight,
You'll see your future shining bright.
    Believe in yourself, for you have the power,
To conquer Trigonometry and make it flower,
Embrace the beauty it holds within,
And let your journey to success begin.

TWENTY-SEVEN

STAY THE COURSE

In the realm of angles and lines,
Where the mysterious unknown resides,
Lies the gateway to cosmic designs,
Where Trigonometry's power abides.

Oh, student, brave and bold,
Embark on this journey untold,
Unlock the secrets of the stars,
And erase your doubts and scars.

Trigonometry, the key to it all,
Unveiling the universe's call,
With triangles and ratios, you shall see,
The beauty of this celestial decree.

Let perseverance be your guiding light,
Through equations and formulas, take flight,
For in the realm of Trigonometry,
Lies the power to shape your destiny.

Fear not the challenges that lie ahead,
For you possess the strength to tread,
Through the depths of angles and degrees,
Unleash your potential, set it free.

With determination as your guiding force,
Conquer the subject, stay the course,
For in the realm of Trigonometry,
A world of possibilities you shall see.

TWENTY-EIGHT

YOUR BRILLIANCE SHINE

In the realm of numbers, you shall find,
A subject that can expand your mind.
Trigonometry, a challenge you face,
But fear not, for you're in the right place.

Angles and triangles, they hold the key,
To unlock the wonders that you seek.
For in this realm of shapes and lines,
The beauty of math truly shines.

With perseverance and dedication, my friend,
You can conquer Trigonometry till the end.
Believe in yourself, have faith in your might,
And you'll rise above any daunting height.

Let the unit circle be your guide,
As you navigate this mathematical tide.

Sine, cosine, tangent, they'll lead the way,
To understanding the universe, night and day.

From the stars that sparkle in the night,
To the waves crashing with all their might,
Trigonometry unveils the secrets untold,
Connecting you to the cosmos, bold.

So embrace the challenge, stand up tall,
For Trigonometry shall never let you fall.
Unlock your potential, let your brilliance shine,
And pass this subject with a victory so divine.

Remember, my friend, the power you hold,
To conquer Trigonometry, brave and bold.
With every equation you solve, you'll find,
A world of knowledge, waiting to unwind.

TWENTY-NINE

SOAR THROUGH THE HEAVENS

In the realm where numbers dance and twirl,
There lies a language, a celestial pearl.
Trigonometry, the gateway to the skies,
Unlocks the secrets that make stars arise.

Embrace the challenge, oh student bold,
Let triangles guide you, their stories unfold.
For in their angles and ratios, you'll find,
The power to shape your brilliant mind.

From sine to cosine, and tangent too,
Equations whisper the universe's clue.
With every calculation, a mystery unravels,
Revealing the cosmos, in all its marvels.

Fear not the symbols, the x's and y's,
For they hold the key to infinite skies.
Let passion be your compass, determination your

guide,
As you navigate this journey, side by side.

    Believe in yourself, and you shall succeed,
For Trigonometry is not a foe, but a need.
Unlock the beauty, embrace the grace,
And watch as your dreams fall into place.

    So, soar through the heavens, reach for the stars,
Trigonometry will carry you far.
With its language divine, you'll shape your destiny,
And find the universe within your reach, you'll see.

THIRTY

DIVE DEEP

In the realm of angles and lines,
Where the cosmic dance intertwines,
There lies the power to see beyond,
To grasp the secrets the universe spawned.
    Oh, student of numbers and forms,
Embrace the challenge that Trigonometry adorns,
For in its depths, lies a treasure untold,
A gateway to knowledge, a story to unfold.
    Like shooting stars across the night,
Trigonometry guides you with its light,
Through triangles and circles, it reveals,
The beauty of patterns, the way life conceals.
    Let your mind soar on the wings of math,
Dive into the depths of its logarithmic path,

For every equation, every theorem you learn,
Unlocks a door, ignites a fire, and makes you yearn.

Persevere, dear student, through the trials you face,
In the realm of Trigonometry, leave your trace,
For with every conquered problem and equation,
You'll find strength, wisdom, and elation.

Believe in yourself, trust in your might,
Embrace the challenge, shine ever so bright,
For Trigonometry is not just a subject to pass,
It's a journey of discovery, a looking glass.

So let the stars guide you, as you dive deep,
Into the mysteries Trigonometry will keep,
Unlock the universe, unravel its mystery,
And soar to heights, never before seen in history.

THIRTY-ONE

SUCCESS WILL SURELY COME TO YOU

In the realm of angles and lines,
Where mathematics intertwines,
Lies the subject of Trigonometry,
A gateway to knowledge and harmony.

Fear not the triangles, my friend,
For in their depths, beauty transcends.
Unlock the secrets they hold within,
And let your journey of learning begin.

Trigonometry, a cosmic dance,
Guiding stars in their celestial trance.
With formulas and ratios, we explore,
The universe's patterns, forevermore.

Embrace the challenge, don't shy away,
For through struggle, we find our way.

In every problem, a chance to grow,
To expand our minds and let them glow.
　Trust in yourself, you have what it takes,
To conquer Trigonometry's vast lakes.
With perseverance and a heart that's true,
Success will surely come to you.
　Let passion guide you through the night,
Illuminate the path with your inner light.
For in understanding this mystical art,
You'll shape your destiny, play your part.
　So rise, young scholar, embrace the quest,
Discover the wonders that lie in Trigonometry's nest.
With determination, you'll soar high,
And touch the realms where dreams can't die.

THIRTY-TWO

STRONG AND TRUE

In the realm where angles dance,
And circles spin their cosmic trance,
There lies a path, both steep and wide,
Where Trigonometry does reside.

    Fear not the triangles, my friend,
For in their angles, knowledge blends,
Unlock the secrets they conceal,
And let your mind begin to heal.

    For Trigonometry, it holds the key,
To galaxies vast and deep blue sea,
It measures stars that light the night,
And guides the sailor in twilight.

    Let not frustration cloud your sight,
Embrace the challenge, hold it tight,

For in the struggle, you shall find,
A strength within, an awakened mind.

Trust in yourself, and you shall see,
The beauty of Trigonometry,
It shapes our world, both near and far,
Unveiling truths like shining stars.

So, let passion be your guiding light,
As you conquer Trigonometry's might,
Believe in your abilities, strong and true,
And let your inner light shine through.

THIRTY-THREE

DETERMINATION AS YOUR LOYAL FRIEND

In the realm of numbers, where angles reside,
Lies the realm of Trigonometry, open wide.
A cosmic dance of triangles and stars,
Where knowledge beckons from afar.

Embrace this realm, young student true,
Let curiosity guide you through.
For in these angles, secrets lie,
Unveiling the mysteries of the sky.

Like waves that crash upon the shore,
Trigonometry's beauty will leave you in awe.
Its formulas and functions, a symphony,
Unraveling the cosmic tapestry.

From the rising sun to the evening moon,
Trigonometry paints the celestial tune.

Unlock the path to the unknown,
And watch the universe's secrets be shown.

Fear not the challenges that lie ahead,
For within you, the power is spread.
Trust in yourself, and you shall find,
The brilliance of Trigonometry, so kind.

So, student, let passion be your guide,
Through the realm where stars reside.
Embrace the numbers, the angles, the art,
And Trigonometry shall ignite your heart.

With determination as your loyal friend,
Success in Trigonometry, you'll transcend.
Believe in yourself, and you shall see,
The wonders of Trigonometry, set yourself free.

THIRTY-FOUR

TAKE A LEAP

In the realm of angles and lines,
A world of mysteries entwines.
Trigonometry, a path untold,
A journey of discovery, bold.
    Embrace the challenge, fear no more,
For within lies the cosmic lore.
Equations dance, formulas sway,
Unlocking secrets, day by day.
    Let passion be your guiding light,
Illuminate the darkest night.
Within the triangles, beauty lies,
A symphony that fills the skies.
    Trust yourself, oh student brave,
With determination, you shall pave

A path to success, where dreams ignite,
And knowledge soars to infinite heights.
   For Trigonometry holds the key,
To understand the world's decree.
From stars that twinkle in the night,
To waves that crash with all their might.
   It's not just numbers on a page,
But a gateway to a boundless stage.
A chance to shape your destiny,
And unravel life's grand tapestry.
   So fear not, dear student, take a leap,
The universe is yours to keep.
Believe in yourself, let passion ignite,
And Trigonometry shall guide you right.

THIRTY-FIVE

TRIGONOMETRY CAN SET YOU FREE

In the realm of angles and lines,
Where triangles entwine,
Lies a subject that holds the key,
To unlocking all that you can be.

Trigonometry, a wondrous art,
That ignites a flame within your heart,
With sin and cos, and tan so bright,
It leads you through the darkest night.

When equations dance upon the page,
And formulas become your stage,
Embrace the challenge with open eyes,
For Trigonometry unveils the skies.

Through radians and degrees we soar,
Exploring realms we can't ignore,

From the heights of mountains to the depths of seas,
Trigonometry unveils the mysteries.
  It's not just numbers, shapes, and graphs,
But a language that the universe speaks,
In every curve and every wave,
Trigonometry reveals its mystique.
  So study hard, my dear friend,
For Trigonometry will transcend,
The boundaries that hold you back,
And guide you on the right track.
  Let the stars be your inspiration,
As you navigate this cosmic equation,
Pass the test and you will see,
The wonders Trigonometry can set you free.

ABOUT THE AUTHOR

Walter the Educator is one of the pseudonyms for Walter Anderson. Formally educated in Chemistry, Business, and Education, he is an educator, an author, a diverse entrepreneur, and he is the son of a disabled war veteran. "Walter the Educator" shares his time between educating and creating. He holds interests and owns several creative projects that entertain, enlighten, enhance, and educate, hoping to inspire and motivate you.

Follow, find new works, and stay up to date
with Walter the Educator™
at WaltertheEducator.com

www.ingramcontent.com/pod-product-compliance
Lightning Source LLC
LaVergne TN
LVHW052000060526
838201LV00059B/3749